In the Sky

The Stars

by Carol Ryback

Reading consultant: Susan Nations, M.Ed., author/literacy coach/ consultant in literacy development

Gareth Stevens
PUBLISHING

Please visit our website, www.garethstevens.com.
For a free color catalog of all our high-quality books, call toll free 1-800-542-2595 or fax 1-877-542-2596.

Library of Congress Cataloging-in-Publication Data

Ryback, Carol.
The stars / by Carol Ryback.
p. cm. — (In the sky)
Includes bibliographical references and index.
ISBN-13: 978-0-8368-6345-1 (lib. bdg.)
ISBN-10: 0-8368-6345-3 (lib. bdg.)
ISBN-13: 978-0-8368-6350-5 (softcover)
ISBN-10: 0-8368-6350-X (softcover)
1. Stars—Juvenile literature. I. Title.
QB801.7.R93 2006
523.8—dc22 2005026536

This edition first published in 2006 by
Gareth Stevens Publishing
111 East 14th Street, Suite 349
New York, NY 10003

Series editor: Dorothy L. Gibbs
Editor: Barbara Kiely Miller
Art direction, cover and layout design: Tammy West
Photo research: Diane Laska-Swanke

Photo credits: Cover, title, pp. 7, 9, 15, 18 NASA and The Hubble Heritage Team, STScI/AURA; p. 5 NOAO; p. 6 ESA/NASA; pp. 10, 20 NASA; p. 11 NOAA; p. 12 NAIC-Arecibo Observatory; p. 14 Tammy West/© Weekly Reader Early Learning Library, 2006; p. 17 NASA, ESA, and The Hubble Heritage Team, STScI; p. 19 NOAO/AURA/NSF; p. 21 NASA/CXC/PSU/S. Park & D. Burrows and NASA/STScI/CfA/P. Challis

Printed in the United States of America

3 4 5 6 7 8 9 10 09 08

Table of Contents

On the cover and title page: Most of the thousands of stars in this giant mass of stars are about three times as old as our Sun.

CHAPTER 1

Starry, Starry Skies

Billions of stars fill the night sky. Did you ever try to count them? If you counted one star every second for the rest of your life, you could never count all the stars!

All stars are not the same. They come in many different sizes, and some shine more brightly than others. A bright star might be a big star. A small star that is close to Earth can also look bright. Our Sun is a medium-size star, but it is so close to Earth that it looks much larger than all the other stars.

Scientists can only guess how many stars are in this bunch, called M4.

Blue gas floats like a veil in space to the side of star cloud NGC 1850.

Stars have no solid parts. They are balls of exploding gas. We see their bright explosions as points of starlight in outer space. Starlight has to travel a long way. It takes a long time to get to Earth. Some stars are so far out in space that their light has been traveling for millions of years and still has not reached Earth!

Earth and the stars are too far apart to measure the distances in miles. Instead, we measure distances between Earth and stars with a unit called a light-year. A **light-year** measures how far a beam of light travels in one year.

Light from these stars started traveling toward us about twelve thousand years ago, when ice covered most of Earth.

CHAPTER 2

A Star Is Born

How does a star form? Far out in space, huge clouds of gas and dust crash into each other. The gas and dust burn and explode. Soon, only gas is left to burn and explode. The explosions make starlight. A star is born!

Most new stars form in areas of space called star nurseries. Star nurseries are far from Earth. Pictures of star nurseries show bright patches and dark patches and lots of colors. Do stars twinkle? Not in space—Earth's atmosphere bends starlight to make it twinkle for us.

A dark cloud of dust in this star nursery looks like the head of a horse.

Starlight from hotter and cooler stars shines in different colors. You might see stars that look red, yellow, white, or blue.

The color of a star shows how hot the star is. All stars are very hot, but some are hotter than others. The light from the hottest stars looks blue or white. The light from cooler stars looks yellow or red. Our Sun looks yellow. It is one of the cooler stars, but it is still very, very hot!

The heat that comes from stars is a type of energy. Heat energy from our star, the Sun, warms Earth's atmosphere and causes our weather. The light that comes from stars is also a kind of energy.

Heat energy from our star, the Sun, sometimes causes a hurricane to form. This photo was taken from space.

A large, dish-shaped instrument in Arecibo, Puerto Rico, listens for noises from the stars.

Did you know that stars make noise? Astronomers use special instruments to listen for noises that come from stars. Star noises sound like grains of sand hitting a bell.

CHAPTER 3

Patterns in the Sky

Long ago, people noticed that some of the stars form patterns in the night sky. They called these patterns of stars **constellations**. They named the constellations after objects, animals, and human figures.

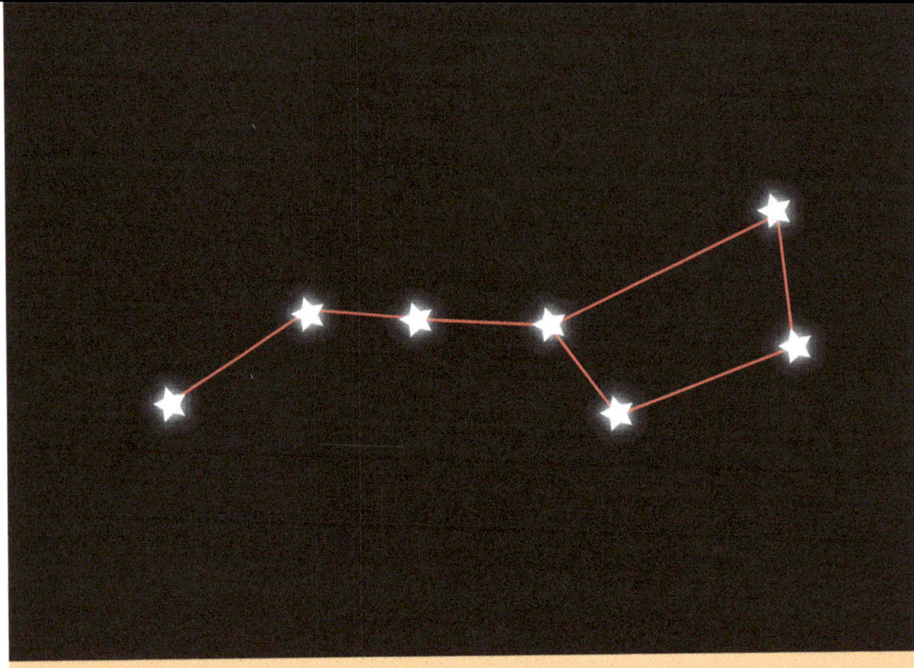

The Big Dipper is easy to find in the summer sky. It is part of a constellation called the Great Bear.

You have probably seen a pattern of stars called the Big Dipper. It looks like a big scoop with a long handle. By itself, the Big Dipper is not a constellation. It is part of a famous constellation known as Ursa Major, or the Great Bear.

Orion is also a famous constellation. You can see Orion in the winter sky. Orion is named after a Greek god known as "the Hunter." Look for three bright stars in a row. They form Orion's belt. Other bright stars mark Orion's shoulders, feet, and sword.

To find Orion, look for the three stars that form its belt.

CHAPTER 4

Galaxies, Pulsars, and Supernovas

Have you ever heard of the Milky Way? The real Milky Way is not a candy bar! It is a gigantic group of stars that is much bigger than any constellation. It looks like a wide path of stars spreading across the sky. It has so many stars that it makes the sky look milky.

The Milky Way is a galaxy. Space holds billions of galaxies. These huge groups of stars form many shapes, including bars, ovals, and pinwheels. Our Sun, its planets, and billions of stars belong to the beautiful Milky Way Galaxy. This galaxy is shaped like a pinwheel.

Like our Milky Way Galaxy, the two galaxies below are also shaped like pinwheels.

Astronomers say that only one of the stars in this huge star group is a blinking star, or "pulsar."

In 1967, an astronomer named Jocelyn Bell Burnell noticed a blinking star. Soon, other astronomers found more blinking stars. Stars blink when they spin superfast. Astronomers call the blinks from a star "pulses." They call blinking stars "pulsars."

Blinking stars are not common. The sky holds billions of stars. So far, astronomers have discovered only about seven hundred blinking stars.

A pulsar lies in the center of a star cloud called the Crab Nebula.

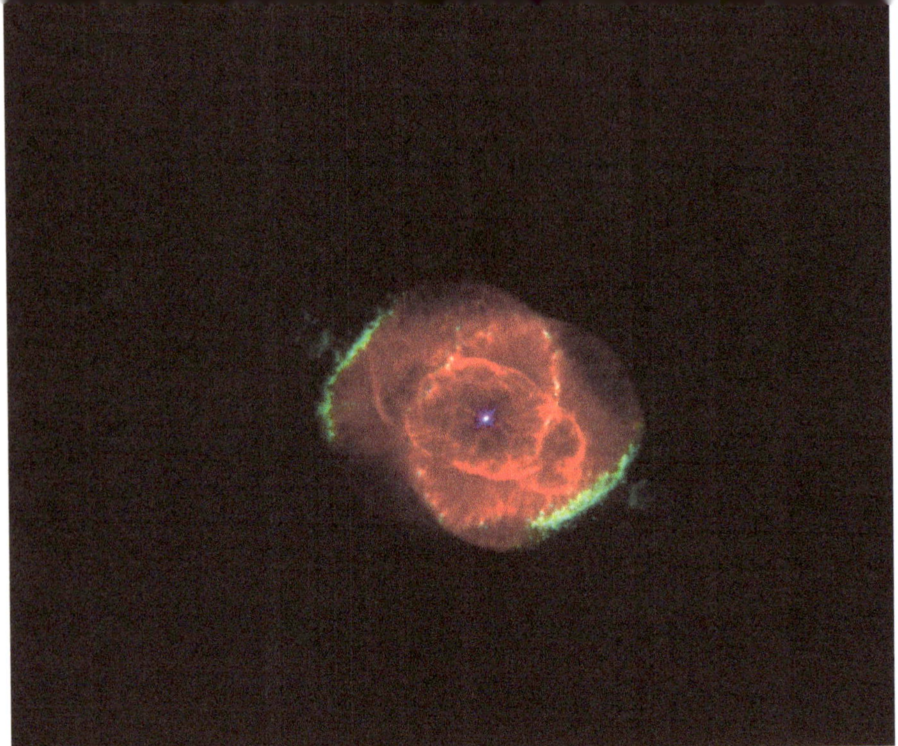

A dying star formed this star cloud, called the Cat's Eye Nebula.

Stars live a long time, but no star shines forever. Stars die. Most dying stars just fade away. Others die in bright blasts called **supernovas.** A supernova can outshine our Sun, even during the day.

The last supernova seen from Earth happened in 1987. It did not happen in the Milky Way Galaxy. It happened in a nearby galaxy. Maybe you will be lucky enough to see the next supernova!

The star that became Supernova 1987A was the size of twenty of our Suns. The explosion lit up a ring of gases in space. Two different instruments took these pictures.

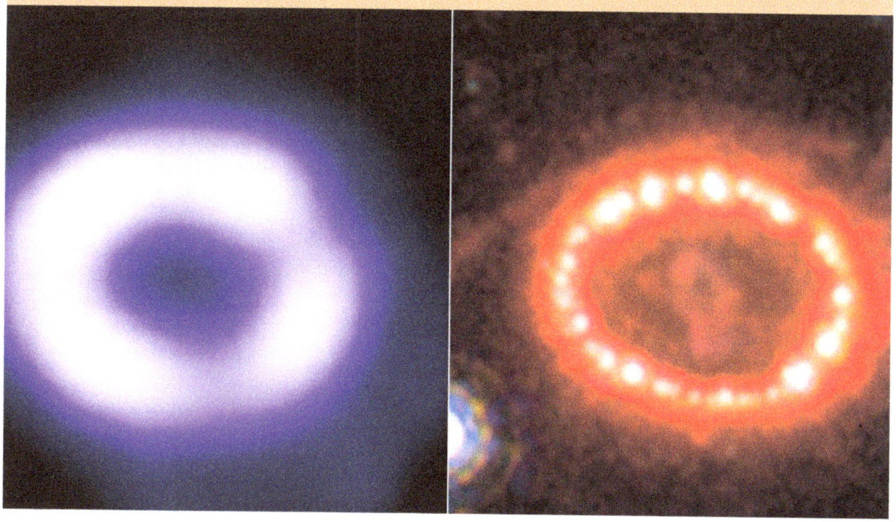

Glossary

astronomer — a scientist who studies stars

atmosphere — the blanket of air around Earth

energy — power that can be used to do physical work

light-year — the distance that light travels in one year

patterns — groups of shapes or objects that look like pictures of things

pulsars — rapidly spinning stars that send out energy in short blinks

supernova — a gigantic explosion of a dying star

For More Information

Books

Galaxies. Seymour Simon (William Morrow & Co. Library)

Once Upon a Starry Night: A Book of Constellations. Jacqueline Mitton (National Geographic Children's Books)

The Stars. Space Explorer (series). Patricia Whitehouse (Heinemann Library)

Watching the Stars. Welcome Books Watching Nature (series). Edana Eckart (Children's Press)

Web Sites

KidsAstronomy.com: Stars

www.kidsastronomy.com/stars.htm

Learn and see the difference between red, yellow, and blue stars and read about groups of stars called dwarfs and giants.

Index

About the Author

Carol Ryback remembers saving her allowance to buy a school binder that featured the planets. She still finds outer space and other "scientific stuff" fascinating. A lifelong Wisconsin resident, Carol's favorite dog stars are golden retrievers Bailey, Merlin, and Harley Taylorson. When not stargazing, Carol likes to scuba dive.